YOUR KNOWLEDGE HAS VALUE

- We will publish your bachelor's and
  master's thesis, essays and papers

- Your own eBook and book -
  sold worldwide in all relevant shops

- Earn money with each sale

Upload your text at www.GRIN.com
and publish for free

C. P. Kumar

Assessment of Groundwater Potential

GRIN Verlag

Bibliografische Information der Deutschen Nationalbibliothek:

Die Deutsche Bibliothek verzeichnet diese Publikation in der Deutschen National-
bibliografie; detaillierte bibliografische Daten sind im Internet über http://dnb.d-
nb.de/ abrufbar.

Imprint:

Copyright © 2014 GRIN Verlag GmbH
Druck und Bindung: Books on Demand GmbH, Norderstedt Germany
ISBN: 978-3-656-75556-2

This book at GRIN:

http://www.grin.com/en/e-book/281601/assessment-of-groundwater-potential

ASSESSMENT OF GROUNDWATER POTENTIAL

C. P. Kumar

Scientist 'F', National Institute of Hydrology, Roorkee – 247667 (Uttarakhand)

1.0   INTRODUCTION

Rapid industrial development, urbanisation and increase in agricultural production have led to freshwater shortages in many parts of the world. In view of increasing demand of water for various purposes like agricultural, domestic and industrial etc., a greater emphasis is being laid for a planned and optimal utilisation of water resources. The water resources of the basins remain almost constant while the demand for water continues to increase. The utilisable water resources of India are estimated to be 1121 BCM out of which 690 BCM is surface water resources and 431 BCM is groundwater resources.

Due to uneven distribution of rainfall both in time and space, the surface water resources are unevenly distributed. Also, increasing intensities of irrigation from surface water alone may result in alarming rise of water table creating problems of water-logging and salinisation, affecting crop growth adversely and rendering large areas unproductive. This has resulted in increased emphasis on development of groundwater resources. The simultaneous development of groundwater, specially through dug wells and shallow tubewells, will lower water table, provide vertical drainage and thus can prevent water-logging and salinisation. Areas, which are already waterlogged, can also be reclaimed.

On the other hand, continuous increased withdrawals from a groundwater reservoir in excess of replenishable recharge may result in regular lowering of water table. In such a situation, a serious problem is created resulting in drying of shallow wells and increase in pumping head for deeper wells and tubewells. This has led to emphasis on planned and optimal development of water resources. An appropriate strategy will be to develop water resources with planning based on conjunctive use of surface water and groundwater.

For sustainable development of water resources, it is imperative to make a quantitative estimation of the available water resources. For this, the first task would be to make a realistic assessment of the surface water and groundwater resources and then plan their use in such a way that full crop water requirements are met and there is neither water-logging nor excessive lowering of groundwater table. It is necessary to maintain the groundwater reservoir in a state of dynamic equilibrium over a period of time and the water level fluctuations have to be kept within a particular range over the monsoon and non-monsoon seasons.

Groundwater is a dynamic system. The total annual replenishable groundwater resource of India is around 431 BCM. Inspite of the national scenario on the availability of groundwater being favourable, there are many areas in the country facing scarcity of water. This is because of the unplanned groundwater development resulting in fall of water levels, failure of wells, and salinity ingress in coastal areas. The development and over-exploitation of groundwater resources in certain parts of the country have raised the concern and need for judicious and scientific resource management and conservation.

The '*National Water Policy*' adopted by the Government of India in 1987 and revised in 2002 and 2012, regards water as one of the most crucial elements in developmental planning. Regarding groundwater, it recommends that

♦ A portion of river flows should be kept aside to meet ecological needs ensuring that the low and high flow releases are proportional to the natural flow regime, including base flow contribution in the low flow season through regulated ground water use.

♦ The anticipated increase in variability in availability of water because of climate change should be dealt with by increasing water storage in its various forms, namely, soil moisture, ponds, ground water, small and large reservoirs and their combination. States should be incentivized to increase water storage capacity, which inter-alia should include revival of traditional water harvesting structures and water bodies.

♦ There is a need to map the aquifers to know the quantum and quality of ground water resources (replenishable as well as non-replenishable) in the country. This process should be fully participatory involving local communities. This may be periodically updated.

♦ Declining ground water levels in over-exploited areas need to be arrested by introducing improved technologies of water use, incentivizing efficient water use and encouraging community based management of aquifers. In addition, where necessary, artificial recharging projects should be undertaken so that extraction is less than the recharge. This would allow the aquifers to provide base flows to the surface system, and maintain ecology.

♦ Water saving in irrigation use is of paramount importance. Recycling of canal seepage water through conjunctive ground water use may also be considered.

♦ There should be concurrent mechanism involving users for monitoring if the water use pattern is causing problems like unacceptable depletion or building up of ground waters, salinity, alkalinity or similar quality problems, etc., with a view to planning appropriate interventions.

♦ The over-drawal of groundwater should be minimized by regulating the use of electricity for its extraction. Separate electric feeders for pumping ground water for agricultural use should be considered.

♦ Quality conservation and improvements are even more important for ground waters, since cleaning up is very difficult. It needs to be ensured that industrial effluents, local cess pools, residues of fertilizers and chemicals, etc., do not reach the ground water.

♦ Industries in water short regions may be allowed to either withdraw only the make up water or should have an obligation to return treated effluent to a specified standard back to the hydrologic system. Tendencies to unnecessarily use more water within the plant to avoid treatment or to pollute ground water need to be prevented.

♦ Appropriate institutional arrangements for each river basin should be developed to collect and collate all data on regular basis with regard to rainfall, river flows, area irrigated by crops and by source, utilizations for various uses by both surface and ground water and to publish water accounts on ten daily basis every year for each river basin with appropriate water budgets and water accounts based on the hydrologic balances. In addition, water budgeting and water accounting should be carried out for each aquifers.

♦ Appropriate institutional arrangements for each river basin should also be developed for monitoring water quality in both surface and ground waters.

A complexity of factors - hydrogeological, hydrological and climatological, control the groundwater occurrence and movement. The precise assessment of recharge and discharge is rather difficult, as no techniques are currently available for their direct measurements. Hence, the methods employed for groundwater resource estimation are all indirect. Groundwater being a dynamic and replenishable resource is generally estimated based on the component of annual recharge, which could be subjected to development by means of suitable groundwater structures.

For quantification of groundwater resources, proper understanding of the behaviour and characteristics of the water bearing rock formation, known as aquifer, is essential. An aquifer has two main functions - (i) to transit water (conduit function) and (ii) to store it (storage function). The groundwater resources in unconfined aquifers can be classified as static and dynamic. The static resources can be defined as the amount of groundwater available in the permeable portion of the aquifer below the zone of water level fluctuation. The dynamic resources can be defined as the amount of groundwater available in the zone of water level fluctuation. The replenishable groundwater resource is essentially a dynamic resource which is replenished annually or periodically by precipitation, irrigation return flow, canal seepage, tank seepage, influent seepage, etc.

The methodologies adopted for computing groundwater resources, are generally based on the hydrologic budget techniques. The hydrologic equation for groundwater regime is a specialized form of water balance equation that requires quantification of the components of inflow to and outflow from a groundwater reservoir, as well as changes in storage therein. Some of these are directly measurable, few may be determined by differences between measured volumes or rates of flow of surface water, and some require indirect methods of estimation.

Water balance techniques have been extensively used to make quantitative estimates of water resources and the impact of man's activities on the hydrological cycle. The study of water balance requires the systematic presentation of data on the water supply and its use within a given study area for a specific period. The water balance of an area is defined by the hydrologic equation, which is basically a statement of the law of conservation of mass as applied to the hydrological cycle. With water balance approach, it is possible to evaluate quantitatively individual contribution of sources of water in the system, over different time periods, and to establish the degree of variation in water regime due to changes in components of the system.

A basinwise approach yields the best results where the groundwater basin can be characterized by prominent drainages. A thorough study of the topography, geology and aquifer conditions should be taken up. The limit of the groundwater basin is controlled not only by topography but also by the disposition, structure and permeability of rocks and the configuration of the water table.

Generally, in igneous and metamorphic rocks, the surface water and groundwater basins are coincident for all practical purposes, but marked differences may be encountered in stratified sedimentary formations. Therefore, the study area for groundwater balance study is preferably taken as a doab which is bounded on two sides by two streams and on the other two sides by other aquifers or extension of the same aquifer. Once the study area is identified, comprehensive studies can be undertaken to estimate for selected period of time, the input and output of water, and change in storage to draw up water balance of the basin.

The estimation of groundwater balance of a region requires quantification of all individual inflows to or outflows from a groundwater system and change in groundwater storage over a given time period. The basic concept of water balance is:

Input to the system - outflow from the system =   change in storage of the system
                                                       (over a period of time)

The general methodology of computing groundwater balance consists of the following:

- Identification of significant components,
- Evaluating and quantifying individual components, and
- Presentation in the form of water balance equation.

The groundwater balance study of an area may serve the following purposes:

- As a check on whether all flow components involved in the system have been quantitatively accounted for, and what components have the greatest bearing on the problem under study.
- To calculate one unknown component of the groundwater balance equation, provided all other components are quantitatively known with sufficient accuracy.
- As a model of the hydrological processes under study, which can be used to predict the effect that changes imposed on certain components will have on the other components of groundwater system.

2.0   GROUNDWATER BALANCE EQUATION

Considering the various inflow and outflow components in a given study area, the groundwater balance equation can be written as:

$$R_r + R_c + R_i + R_t + S_i + I_g = E_t + T_p + S_e + O_g + \Delta S \qquad \text{...(1)}$$

where,

$$
\begin{aligned}
R_r &= \text{recharge from rainfall;} \\
R_c &= \text{recharge from canal seepage;} \\
R_i &= \text{recharge from field irrigation;} \\
R_t &= \text{recharge from tanks;} \\
S_i &= \text{influent seepage from rivers;} \\
I_g &= \text{inflow from other basins;} \\
E_t &= \text{evapotranspiration from groundwater;} \\
T_p &= \text{draft from groundwater;} \\
S_e &= \text{effluent seepage to rivers;} \\
O_g &= \text{outflow to other basins; and}
\end{aligned}
$$

ΔS =   change in groundwater storage.

Preferably, all elements of the groundwater balance equation should be computed using independent methods. However, it is not always possible to compute all individual components of the groundwater balance equation separately. Sometimes, depending on the problem, some components can be lumped, and account only for their net value in the equation.

Computations of various components usually involve errors, due to shortcomings in the estimation techniques. The groundwater balance equation therefore generally does not balance, even if all its components are computed by independent methods. The resultant discrepancy in groundwater balance is defined as a residual term in the balance equation, which includes errors in the quantitative determination of various components as well as values of the components which have not been accounted in the equation.

The water balance may be computed for any time interval. The complexity of the computation of the water balance tends to increase with increase in area. This is due to a related increase in the technical difficulty of accurately computing the numerous important water balance components.

3.0   DATA REQUIREMENTS FOR A GROUNDWATER BALANCE STUDY

For carrying out a groundwater balance study, following data may be required over a given time period:

Rainfall data: Monthly rainfall data of sufficient number of rainguage stations lying within or around the study area, along with their locations, should be available.

Land use data and cropping patterns: Land use data are required for estimating the evapotranspiration losses from the water table through forested area. Cropping pattern data are necessary for estimating the spatial and temporal distributions of groundwater withdrawals, if required. Monthly pan evaporation rates should also be available at few locations for estimation of consumptive use requirements of different crops.

River data: Monthly river stage and discharge data along with river cross-sections are required at few locations for estimating the river-aquifer interflows.

Canal data: Monthwise water releases into the canal and its distributaries along with running days during each month are required. To account for the seepage losses through the canal system, the seepage loss test data are required in different canal reaches and distributaries.

Tank data: Monthly tank gauges and water releases should be available. In addition, depth vs. area and depth vs. capacity curves should also be available for computing the evaporation and seepage losses from tanks. Field test data are required for computing infiltration capacity to be used to evaluate the recharge from depression storage.

Water table data: Monthly water table data (or at least pre-monsoon and post-monsoon data) from sufficient number of well-distributed observation wells along with their locations are required. The available data should comprise reduced level (R.L.) of water table and depth to water table.

Groundwater draft: For estimating groundwater withdrawals, the number of each type of wells operating in the area, their corresponding running hours each month and discharge are required. If a complete inventory of wells is not available, then this can be obtained by carrying out sample surveys.

Aquifer parameters: Data regarding the storage coefficient and transmissivity are required at sufficient number of locations in the study area.

4.0   GROUNDWATER RESOURCE ESTIMATION METHODOLOGY

The Groundwater Estimation Committee (GEC) was constituted by the Government of India in 1982 to recommend methodologies for estimation of the groundwater resource potential in India. It was recommended by the committee that the groundwater recharge should be estimated based on groundwater level fluctuation method. However, in areas, where groundwater level monitoring is not being done regularly, or where adequate data about groundwater level fluctuation is not available, adhoc norms of rainfall infiltration may be adopted. In order to review the recommended methodology, the committee was reconstituted in 1995, which released its report in 1997. This committee proposed several improvements in the existing methodology based on groundwater level fluctuation approach. Salient features of their recommendations are given below.

(a)   Watershed may be used as the unit for groundwater resource assessment in hard rock areas, which occupies around $2/3^{rd}$ part of the country. The size of the watershed as a hydrological unit could be of about 100 to 300 sq. km. area. The assessment made for watershed as unit may be transferred to administrative unit such as block, for planning development programmes.

(b)   For alluvial areas, the present practice of assessment based on block/taluka/mandal-wise basis is retained. The possibility of adopting doab as the unit of assessment in alluvial areas needs further detailed studies.

(c)   The total geographical area of the unit for resource assessment is to be divided into subareas such as hilly regions (slope > 20%), saline groundwater areas, canal command areas and non-command areas, and separate resource assessment may be made for these subareas. Variations in geomorphological and hydrogeological characteristics may be considered within the unit.

(d)   For hard rock areas, the specific yield value may be estimated by applying the water level fluctuation method for the dry season data, and then using this specific yield value in the water level fluctuation method for the monsoon season to get recharge. For alluvial areas, specific yield values may be estimated from analysis of pumping tests. However, norms for specific yield values in different hydrogeological regions may still be necessary for use in situations where the above methods are not feasible due to inadequacy of data.

(e)   There should be at least 3 spatially well-distributed observation wells in the unit, or one observation well per 100 sq. km. whichever is more.

(f)   The problem of accounting for groundwater inflow/outflow and base flow from a region is difficult to solve. If watershed is used as a unit for resource assessment in hard rock areas, the groundwater inflow/outflow may become negligible. The base flow can be estimated if one stream gauging station is located at the exit of the watershed.

(g)   Norms for return flow from groundwater and surface water irrigation are revised taking into account the source of water (groundwater/surface water), type of crop (paddy/non-paddy) and depth of groundwater level.

In subsequent section, the recommended GEC norms for estimation of various inflow/outflow components of the groundwater balance equation have been mentioned at appropriate places, along with other methodologies/formulae in use.

5.0   ESTIMATION OF GROUNDWATER BALANCE COMPONENTS

The various inflow/outflow components of the groundwater balance equation may be estimated through appropriate empirical relationships suitable for a region, Groundwater Estimation Committee norms (1997), field experiments or other methods, as discussed below.

5.1   Recharge from Rainfall (R_r)

Rainfall is the major source of recharge to groundwater. Part of the rain water, which falls on the ground, is infiltrated into the soil. A part of this infiltrated water is utilized in filling the soil moisture deficiency while the remaining portion percolates down to reach the water table, which is termed as *rainfall recharge* to the aquifer. The amount of rainfall recharge depends on various hydrometeorological and topographic factors, soil characteristics and depth to water table. The methods for estimation of rainfall recharge involve the empirical relationships established between recharge and rainfall developed for different regions, Groundwater Resource Estimation Committee norms, groundwater balance approach, and soil moisture data based methods.

5.1.1 Empirical Methods

Several empirical formulae have been worked out for various regions in India on the basis of detailed studies. Some of the commonly used formulae are:

(a) Chaturvedi formula: Based on the water level fluctuations and rainfall amounts in Ganga-Yamuna doab, Chaturvedi in 1936, derived an empirical relationship to arrive at the recharge as a function of annual precipitation.

$$R_r = 2.0 \, (P - 15)^{0.4} \qquad \qquad \dots(2)$$

where,
R_r = net recharge due to precipitation during the year, in inches; and
P = annual precipitation, in inches.

This formula was later modified by further work at the U.P. Irrigation Research Institute, Roorkee and the modified form of the formula is

$$R_r = 1.35 \, (P - 14)^{0.5} \qquad \qquad \dots(3)$$

The Chaturvedi formula has been widely used for preliminary estimations of groundwater recharge due to rainfall. It may be noted that there is a lower limit of the rainfall below which the recharge due to rainfall is zero. The percentage of rainfall recharged commences from zero at P = 14 inches, increases upto 18% at P = 28 inches, and again decreases. The lower limit of rainfall in the formula may account for the soil moisture deficit, the interception losses and potential evaporation. These factors being site specific, one generalized formula may not be applicable to all the alluvial areas. Tritium tracer studies on groundwater recharge in the alluvial deposits of Indo-Gangetic plains of western U.P., Punjab, Haryana and alluvium in Gujarat state have indicated variations with respect to Chaturvedi formula.

(b) Kumar and Seethapathi (2002): They conducted a detailed seasonal groundwater balance study in Upper Ganga Canal command area for the period 1972-73 to 1983-84 to determine groundwater recharge from rainfall. It was observed that as the rainfall increases, the quantity of recharge also increases but the increase is not linearly proportional. The recharge coefficient (based upon the rainfall in monsoon season) was found to vary between 0.05 to 0.19 for the study area. The following empirical relationship (similar to Chaturvedi formula) was derived by fitting the estimated values of rainfall recharge and the corresponding values of rainfall in the monsoon season through the non-linear regression technique.

$$R_r = 0.63 (P - 15.28)^{0.76} \qquad \qquad ...(4)$$

where,

R_r = Groundwater recharge from rainfall in monsoon season (inch);
P = Mean rainfall in monsoon season (inch).

The relative errors (%) in the estimation of rainfall recharge computed from the proposed empirical relationship was compared with groundwater balance study. In almost all the years, the relative error was found to be less than 8%. On the other hand, relative errors (%) computed from Chaturvedi formula (equations 2 and 3) were found to be quite high. Therefore, equation (4) can conveniently be used for better and quick assessment of natural groundwater recharge in Upper Ganga Canal command area.

(c) Amritsar formula: Using regression analysis for certain doabs in Punjab, the Irrigation and Power Research Institute, Amritsar, developed the following formula in 1973.

$$R_r = 2.5 (P - 16)^{0.5} \qquad \qquad ...(5)$$

where, R_r and P are measured in inches.

(d) Krishna Rao: Krishna Rao gave the following empirical relationship in 1970 to determine the groundwater recharge in limited climatological homogeneous areas:

$$R_r = K (P - X) \qquad \qquad ...(6)$$

The following relation is stated to hold good for different parts of Karnataka:

R_r = 0.20 (P - 400) for areas with annual normal rainfall (P) between 400 and 600 mm
R_r = 0.25 (P - 400) for areas with P between 600 and 1000 mm
R_r = 0.35 (P - 600) for areas with P above 2000 mm

where, R_r and P are expressed in millimetres.

The relationships indicated above, which were tentatively proposed for specific hydrogeological conditions, have to be examined and established or suitably altered for application to other areas.

5.1.2 Groundwater Resource Estimation Committee Norms

If adequate data of groundwater levels are not available, rainfall recharge may be estimated using the rainfall infiltration method. The same recharge factor may be used for both monsoon and non-monsoon rainfall, with the condition that the recharge due to non-monsoon rainfall may be taken as zero, if the rainfall during non-monsoon season is less than 10% of annual rainfall. Groundwater Resource Estimation Committee (1997) recommended the following rainfall infiltration factors:

(a) **Alluvial areas**

Indo-Gangetic and inland areas	-	22 %
East coast	-	16 %
West coast	-	10 %

(b) **Hard rock areas**

Weathered granite, gneiss and schist with low clay content	-	11 %
Weathered granite, gneiss and schist with significant clay content	-	8 %
Granulite facies like charnockite etc.	-	5 %
Vesicular and jointed basalt	-	13 %
Weathered basalt	-	7 %
Laterite	-	7 %
Semi-consolidated sandstone	-	12 %
Consolidated sandstone, Quartzites, Limestone (except cavernous limestone)	-	6 %
Phyllites, Shales	-	4 %
Massive poorly fractured rock	-	1 %

An additional 2% of rainfall recharge factor may be used in areas where watershed development with associated soil conservation measures is implemented. This additional factor is separate from contribution due to water conservation structures such as check dams, nalla bunds, percolation tanks etc., for which the norms are defined separately.

5.1.3 Groundwater Balance Approach

In this method, all components of the groundwater balance equation (1), except the rainfall recharge, are estimated individually. The algebraic sum of all input and output components is equated to the change in groundwater storage, as reflected by the water table fluctuation, which in turn yields the single unknown in the equation, namely, the rainfall recharge. A pre-requisite for successful application of this technique is the availability of very extensive and accurate hydrological and meteorological data. The groundwater balance approach is valid for the areas where the year can be divided into monsoon and non-monsoon seasons with the bulk of rainfall occurring in former.

Groundwater balance study for monsoon and non-monsoon periods is carried out separately. The former yields an estimate of recharge coefficient and the later determines the degree of accuracy with which the components of water balance equation have been estimated. Alternatively, the average specific yield in the zone of fluctuation can be determined from a groundwater balance study for the non-monsoon period and using this specific yield, the recharge due to rainfall can be determined using the groundwater balance components for the monsoon period.

5.1.4 Soil Moisture Data Based Methods

Soil moisture data based methods are the lumped and distributed model and the nuclear methods. In the lumped model, the variation of soil moisture content in the vertical direction is ignored and any effective input into the soil is assumed to increase the soil moisture content uniformly. Recharge is calculated as the remainder when losses, identified in the form of runoff and evapotranspiration, have been deducted from the precipitation with proper accounting of soil moisture deficit. In the distributed model, variation of soil moisture content in the vertical direction is accounted and the method involves the numerical solution of partial differential equation (Richards equation) governing one-dimensional flow through unsaturated medium, with appropriate initial and boundary conditions.

(a)   Soil Water Balance Method

Water balance models were developed in the 1940s by Thornthwaite (1948) and revised by Thornthwaite and Mather (1955). The method is essentially a book-keeping procedure which estimates the balance between the inflow and outflow of water. When applying this method to estimate the recharge for a catchment area, the calculation should be repeated for areas with different precipitation, evapotranspiration, crop type and soil type. The soil water balance method is of limited practical value, because evapotranspiration is not directly measurable. Moreover, storage of moisture in the unsaturated zone and the rates of infiltration along the various possible routes to the aquifer form important and uncertain factors. Another aspect that deserves attention is the depth of the root zone which may vary in semi-arid regions between 1 and 30 meters. Results from this model are of very limited value without calibration and validation, because of the substantial uncertainty in input data.

(b)   Nuclear Methods

Nuclear techniques can be used for the determination of recharge by measuring the travel of moisture through a soil column. The technique is based upon the existence of a linear relation between neutron count rate and moisture content (% by volume) for the range of moisture contents generally occurring in the unsaturated soil zone.

Neutron soil moisture probe

Neutron soil moisture probe is used to determine soil moisture contents at different depths in the unsaturated zone. The neutron probe consists of a fast neutron source of radioactive isotopes (Radium and Beryllium) and a slow neutron detector adjacent to the neutron source. The neutron source of the probe emits fast neutrons into the surrounding soil. In soil, hydrogen is present mainly in the form of moisture. Hence, fast neutrons when projected into the soil, are scattered by the hydrogen nuclei after collision and are slowed down to the thermal energy level by losing their kinetic energy. The number of neutrons slowed down thus is proportional to the hydrogen atoms available in the soil and hence, to the moisture content of the soil. These slow down neutrons are detected by a slow neutron detector (thermal neutron detector). The measurement of number of counts per second gives the measure of the soil moisture content, which is converted into actual values of soil moisture content after applying suitable calibration equation. By repeated measurements of moisture contents through neutron soil moisture probe, at different times during the rainy season, groundwater recharge can be estimated.

Tritium tagging technique

The basic principle of estimation of recharge to groundwater by using this method assumes that soil water in the unsaturated zone moves downward 'layer by layer' similar to a piston flow, i.e. if any amount of water is added to the ground surface due to precipitation or irrigation, it will percolate by pushing equal amount of water beneath it further downwards such that an equal amount of the moisture of the last layer in the unsaturated zone is added to the groundwater. If the tritium radioisotope is tagged below the active root zone and sun-heating zone, it will be mixed with the soil moisture available at that depth and act as an impermeable layer. Therefore, if any water is added to the top soil, it will be infiltrated into the soil strata by pushing down the older water, this shift in the tritium can be observed after a specified period depending upon the input supply of water. However, the injected tritium will be found in the form of a broadened peak due to molecular diffusion, stream line dispersion, asymmetrical flow and other heterogeneities of the soil media.

5.2   Recharge from Canal Seepage (R_c)

Seepage refers to the process of water movement from a canal into and through the bed and wall material. Seepage losses from irrigation canals often constitute a significant part of the total recharge to groundwater system. Hence, it is important to properly estimate these losses for recharge assessment to groundwater system. Recharge by seepage from canals depends upon the size and cross-section of the canal, depth of flow, characteristics of soils in the bed and sides, and location as well as level of drains on either side of the canal. A number of empirical formulae and formulae based on theoretical considerations have been proposed to estimate the seepage losses from canals.

Recharge from canals that are in direct hydraulic connection with a phreatic aquifer underlain by a horizontal impermeable layer at shallow depth, can be determined by Darcy's equation, provided the flow satisfies Dupuit assumptions.

$$R_c \;=\; K \,\frac{h_s - h_l}{L}\, A \qquad\qquad\qquad ...(7)$$

where, h_s and h_l are water-level elevations above the impermeable base, respectively, at the canal, and at distance L from it. For calculating the area of flow cross-section, the average of the saturated thickness $(h_s + h_l)/2$ is taken. The crux of computation of seepage depends on correct assessment of the hydraulic conductivity, K. Knowing the percentage of sand, silt and clay, the hydraulic conductivity of undisturbed soil can be approximately determined using the soil classification triangle showing relation of hydraulic conductivity to texture for undisturbed sample (Johnson, 1963).

A number of investigations have been carried out to study the seepage losses from canals. The following formulae/values are in vogue for the estimation of seepage losses:

(a)   As reported by the Indian Standard (IS: 9452 Part 1, 1980), the loss of water by seepage from unlined canals in India varies from 0.3 to 7.0 cumec per million square meter of wetted area depending on the permeability of soil through which the canal passes, location of water table, distance of drainage, bed width, side slope, and depth of water in the canal. Transmission loss of 0.60 cumec per million square meter of wetted area of lined canal is generally assumed (IS: 10430, 1982).

(b)   For unlined channels in Uttar Pradesh, it has been proposed that the losses per million square meter of wetted area are 2.5 cumec for ordinary clay loam to about 5 cumec for sandy loam with an average of 3 cumec. Empirically, the seepage losses can be computed using the following formula:

$$Losses\;in\;cumecs\,/\,km \;=\; \frac{C}{200}\,(B + D)^{2/3} \qquad\qquad ...(8)$$

where, B and D are the bed width and depth, respectively, of the channel in meters, C is a constant with a value of 1.0 for intermittent running channels and 0.75 for continuous running channels.

(c)   For lined channels in Punjab, the following formula is used for estimation of seepage losses:

$$R_c \;=\; 1.25\,Q^{0.56} \qquad\qquad\qquad ...(9)$$

where, R_c is the seepage loss in cusec per million square foot of wetted area and Q, in cusec, is the discharge carried by the channel. In unlined channels, the loss rate on an average is four times the value computed using the above formula.

(d)   U. S. B. R. recommended the channel losses based on the channel bed material as given below:

Material	Seepage Losses (cumec per million square meter of wetted area)

Clay and clay loam	:	1.50
Sandy loam	:	2.40
Sandy and gravely soil	:	8.03
Concrete lining	:	1.20

(e) Groundwater Resource Estimation Committee (1997) has recommended the following norms:

    (i)   Unlined canals in normal soil with some clay content along with sand
            - 1.8 to 2.5 cumec per million square meter of wetted area.

    (ii)   Unlined canals in sandy soil with some silt content
            - 3.0 to 3.5 cumec per million square meter of wetted area.

    (iii) Lined canals and canals in hard rock areas
            - 20% of the above values for unlined canals.

These values are valid if the water table is relatively deep. In shallow water table and water logged areas, the recharge from canal seepage may be suitably reduced. Specific results from case studies may be used, if available. The above norms take into consideration the type of soil in which the canal runs while computing seepage. However, the actual seepage will also be controlled by the width of canal (B), depth of flow (D), hydraulic conductivity of the bed material (K) and depth to water table.

Knowing the values of B and D, the range of seepage losses (R_{c_max} and R_{c_min}) from the canal may be obtained as

$$R_{c\_max} = K(B + 2D) \quad \text{(in case of deeper water table)} \quad \text{...(10a)}$$

$$R_{c\_min} = K(B - 2D) \quad \text{(in case of water table at the level of channel bed)} \quad \text{...(10b)}$$

However, the various guidelines for estimating losses in the canal system, as given above, are at best approximate. Thus, the seepage losses may best be estimated by conducting actual tests in the field. The methods most commonly adopted are:

Inflow - outflow method: In this method, the water that flows into and out of the section of canal, under study, is measured using current meter or Parshall flume method. The difference between the quantities of water flowing into and out of the canal reach is attributed to seepage. This method is advantageous when seepage losses are to be measured in long canal reaches with few diversions.

Ponding method: In this method, bunds are constructed in the canal at two locations, one upstream and the other downstream of the reach of canal with water filled in it. The total change in storage in the reach is measured over a period of time by measuring the rate of drop of water surface elevation in the canal reach. Alternatively, water may be added to maintain a constant water surface elevation. In this case, the volume of water added is measured along with the elapsed time to compute the rate of seepage loss. The ponding method provides an accurate means of measuring seepage losses and is especially suitable when they are small (e.g. in lined canals).

Seepage meter method: The seepage meter is a modified version of permeameter developed for use under water. Various types of seepage meters have been developed. The two most important are seepage meter with submerged flexible water bag and falling head seepage meter. Seepage meters are suitable for measuring local seepage rates in canals or ponds and used only in unlined or earth-lined canals. They are quickly and easily installed and give reasonably satisfactory results for the conditions at the test site but it is difficult to obtain accurate results when seepage losses are low.

The total losses from the canal system generally consist of the evaporation losses (E_c) and the seepage losses (R_c). The evaporation losses are generally 10 to 15 percent of the total losses. Thus the R_c value is 85 to 90 percent of the losses from the canal system.

5.3   Recharge from Field Irrigation (R_i)

Water requirements of crops are met, in parts, by rainfall, contribution of moisture from the soil profile, and applied irrigation water. A part of the water applied to irrigated field crops is lost in consumptive use and the balance infiltrates to recharge the groundwater. The process of re-entry of a part of the water used for irrigation is called return flow. Percolation from applied irrigation water, derived both from surface water and groundwater sources, constitutes one of the major components of groundwater recharge. The irrigation return flow depends on the soil type, irrigation practice and type of crop. Therefore, irrigation return flows are site specific and will vary from one region to another.

For a correct assessment of the quantum of recharge by applied irrigation, studies are required to be carried out on experimental plots under different crops in different seasonal conditions. The method of estimation comprises application of the water balance equation involving input and output of water in experimental fields.

The recharge due to irrigation return flow may also be estimated, based on the source of irrigation (groundwater or surface water), the type of crop (paddy, non-paddy) and the depth of water table below ground surface, using the norms provided by Groundwater Resource Estimation Committee (1997), as given below (as percentage of water application):

Source of Irrigation	Type of Crop	Water table below ground surface		
		<10m	10-25m	>25m
Groundwater	Non-paddy	25	15	5
Surface water	Non-paddy	30	20	10
Groundwater	Paddy	45	35	20
Surface water	Paddy	50	40	25

For surface water, the recharge is to be estimated based on water released at the outlet from the canal/distribution system. For groundwater, the recharge is to be estimated based on gross draft. Where continuous supply is used instead of rotational supply, an additional recharge of 5% of application may be used. Specific results from case studies may be used, if available.

5.4   Recharge from Tanks (R_t)

Studies have indicated that seepage from tanks varies from 9 to 20 percent of their live storage capacity. However, as data on live storage capacity of large number of tanks may not be available, seepage from tanks may be taken as 44 to 60 cm per year over the total water spread, taking into account the agro-climatic conditions in the area. The seepage from percolation tanks is higher and may be taken as 50 percent of its gross storage. In case of seepage from ponds and lakes, the norms as applied to tanks may be taken. Groundwater Resource Estimation Committee (1997) has recommended that based on the average area of water spread, the recharge from storage tanks and ponds may be taken as 1.4 mm/day for the period in which tank has water. If data on the average area of water spread is not available, 60% of the maximum water spread area may be used instead of average area of water spread.

In case of percolation tanks, recharge may be taken as 50% of gross storage, considering the number of fillings, with half of this recharge occurring in monsoon season and the balance in non-monsoon season. Recharge due to check dams and nala bunds may be taken as 50% of gross storage (assuming annual desilting maintenance exists) with half of this recharge occurring in the monsoon season and the balance in the non-monsoon season.

5.5   Influent and Effluent Seepage (S_i & S_e)

The river-aquifer interaction depends on the transmissivity of the aquifer system and the gradient of water table in respect to the river stage. Depending on the water level in the river and in the aquifer (in the vicinity of river), the river may recharge the aquifer (influent) or the aquifer may contribute to the river flow (effluent). The effluent or influent character of the river may vary from season to season and from reach to reach. The seepage from/to the river can be determined by dividing the river reach into small sub-reaches and observing the discharges at the two ends of the sub-reach along with the discharges of its tributaries and diversions, if any. The discharge at the downstream end is expressed as:

$$Q_d \cdot \Delta t = Q_u \cdot \Delta t + Q_g \cdot \Delta t + Q_t \cdot \Delta t - Q_o \cdot \Delta t - E \cdot \Delta t \pm S_{rb} \qquad \ldots(11)$$

where,

Q_d	=	discharge at the downstream section;
Q_u	=	discharge at the upstream section;
Q_g	=	groundwater contribution (unknown quantity; -ve computed value indicates influent conditions);
Q_t	=	discharge of tributaries;
Q_o	=	discharge diverted from the river;
E	=	rate of evaporation from river water surface and flood plain (for extensive bodies of surface water and for long time periods, evaporation from open water surfaces can not be neglected);
S_{rb}	=	change in bank storage (+ for decrease and - for increase); and
Δt	=	time period.

The change in bank storage can be determined by monitoring the water table along the cross-section normal to the river. Thus, using the above equation, seepage from/to the river over a certain period of time Δt can be computed. However, this would be the contribution from aquifers on both sides of the stream. The contribution from each side can be separated by the following method:

$$Contribution\ from\ left\ bank\ =\ \frac{I_L T_L}{I_L T_L + I_R T_R} \cdot Q_g \qquad \qquad \text{...(12)}$$

$$Contribution\ from\ right\ bank\ =\ \frac{I_R T_R}{I_L T_L + I_R T_R} \cdot Q_g \qquad \qquad \text{...(13)}$$

where, I_L and T_L are gradient and transmissivity respectively on the left side and I_R and T_R are those on the right.

5.6 Inflow from and Outflow to Other Basins (I_g & O_g)

For the estimation of groundwater inflow/outflow from/to other basins, regional water table contour maps are drawn based on the observed water level data from wells located within and outside the study area. The flows into and out of a region are governed mainly by the hydraulic gradient and transmissivity of the aquifer. The gradient can be determined by taking the slope of the water table normal to water table contour. The length of the section, across which groundwater inflow/outflow occurs, is determined from contour maps, the length being measured parallel to the contour. The inflow/outflow is determined as follows:

$$I_g\ or\ O_g\ =\ \sum^{L} T\ I\ \Delta L \qquad \qquad \text{...(14)}$$

where, T is the transmissivity and I is the hydraulic gradient averaged over a length ΔL of contour line.

5.7 Evapotranspiration from Groundwater (E_t)

Evapotranspiration is the combined process of transpiration from vegetation and evaporation from both soil and free water surfaces. Potential evapotranspiration is the maximum loss of water through evapotranspiration. Evapotranspiration from groundwater occurs in waterlogged areas or in forested areas with roots extending to the water table. From the land use data, area under forests is available while the waterlogged areas may be demarcated from depth to water table maps. The potential evapotranspiration from such areas can be computed using standard methods.

Depth to water table maps may be prepared based on well inventory data to bring into focus the extensiveness of shallow water table areas. During well inventory, investigation should be specifically oriented towards accurately delineating water table depth for depths less than 2 meters. The evapotranspiration can be estimated based on the following equations:

$$E_t\ =\ PE_t * A \qquad\qquad if\ h > h_s \qquad\qquad \text{...(15a)}$$
$$E_t\ =\ 0 \qquad\qquad if\ h < (h_s - d) \qquad\qquad \text{...(15b)}$$
$$E_t\ =\ PE_t * A\ (h - (h_s - d))/d \qquad\qquad if\ (h_s - d) \leq h \leq h_s \qquad\qquad \text{...(15c)}$$

where,

E_t = evapotranspiration in volume of water per unit time $[L^3\,T^{-1}]$;

PE_t = maximum rate of evapotranspiration in volume of water per unit area per unit time $[L^3\,L^{-2}\,T^{-1}]$;

A = surface area $[L^2]$;

h = water table elevation $[L]$;

h_s = water table elevation at which the evapotranspiration loss reaches the maximum value; and

d = extinction depth. When the distance between h_s and h exceeds d, evapotranspiration from groundwater ceases $[L]$.

5.8 Draft from Groundwater (T_p)

Draft is the amount of water lifted from the aquifer by means of various lifting devices. To estimate groundwater draft, an inventory of wells and a sample survey of groundwater draft from various types of wells (state tubewells, private tubewells and open wells) are required. For state tubewells, information about their number, running hours per day, discharge, and number of days of operation in a season is generally available in the concerned departments. To compute the draft from private tubewells, pumping sets and rahats etc., sample surveys have to be conducted regarding their number, discharge and withdrawals over the season.

In areas where wells are energised, the draft may be computed using power consumption data. By conducting tests on wells, the average draft per unit of electricity consumed can be determined for different ranges in depth to water levels. By noting the depth to water level at each distribution point and multiplying the average draft value with the number of units of electricity consumed, the draft at each point can be computed for every month.

In the absence of sample surveys, the draft can be indirectly estimated from the net crop water requirement which is based upon the cropping pattern and irrigated areas under various crops. The consumptive use requirements of crops are calculated using the consumptive use coefficient and effective rainfall. The consumptive use coefficient for crops is related to percentage of crop growing season (Table 1). The consumptive use for each month can be evaluated by multiplying consumptive use coefficient with monthly pan evaporation rates. For the computation of net irrigation requirement, the effective rainfall has to be evaluated. Effective rainfall is the portion of rainfall that builds up the soil moisture in the root zone after accounting for direct runoff and deep percolation. The normal monthly effective rainfall, as related to average monthly consumptive use, is given in Table 2. Net crop water requirement is obtained after subtracting effective rainfall from consumptive use requirement. The groundwater draft can thus be estimated by subtracting canal water released for the crops from the net crop water requirement.

Table 1: Crop Consumptive Use Coefficients

Percent of crop growing season	Consumptive use (Evapotranspiration) coefficient (to be multiplied by Class 'A' Pan Evaporation)							
	Group A	Group B	Group C	Group D	Group E	Group F	Group G	Rice
0	0.20	0.15	0.12	0.08	0.90	0.60	0.50	0.80
5	0.20	0.15	0.12	0.08	0.90	0.60	0.55	0.90
10	0.36	0.27	0.22	0.25	0.90	0.60	0.60	0.95
15	0.50	0.38	0.30	0.19	0.90	0.60	0.65	1.00
20	0.64	0.48	0.38	0.27	0.90	0.60	0.70	1.05
25	0.75	0.56	0.45	0.33	0.90	0.60	0.75	1.10
30	0.84	0.63	0.50	0.40	0.90	0.60	0.80	1.14
35	0.92	0.69	0.55	0.46	0.90	0.60	0.86	1.17
40	0.97	0.73	0.58	0.52	0.90	0.60	0.90	1.21
45	0.99	0.74	0.60	0.58	0.90	0.60	0.95	1.25
50	1.00	0.75	0.60	0.65	0.90	0.60	1.00	1.30
55	1.00	0.75	0.60	0.71	0.90	0.60	1.00	1.30
60	0.99	0.74	0.60	0.77	0.90	0.60	1.00	1.30
65	0.96	0.72	0.58	0.82	0.90	0.60	0.95	1.25
70	0.91	0.68	0.55	0.88	0.90	0.60	0.90	1.20
75	0.85	0.64	0.51	0.90	0.90	0.60	0.85	1.15
80	0.75	0.56	0.45	0.90	0.90	0.60	0.80	1.00
85	0.60	0.45	0.36	0.80	0.90	0.60	0.75	1.00
90	0.46	0.35	0.28	0.70	0.90	0.60	0.70	0.90
100	0.20	0.20	0.17	0.20	0.90	0.60	0.50	0.20

Group A -   beans, maize, cotton, potatoes, sugar beets, jowar and peas
Group B -   dates, olives, walnuts, tomatoes, and hybrid jowar
Group C -   melons, onions, carrots, and grapes
Group D -   barley, flex, wheat, and other small grains
Group E -   pastures, orchards, and cover crops
Group F -   citrus crops, oranges, limes and grape fruit
Group G -   sugarcane and alfalfa

Table 2: **Normal Monthly Effective Rainfall as related to Normal Monthly Rainfall and Average Monthly Consumptive Use**

Normal Monthly Rainfall (mm)	Average Monthly Consumptive Use (mm)													
	25	50	75	100	125	150	175	200	225	250	275	300	325	350
	Normal Monthly Effective Rainfall (mm)													
25	15	17	18	18	19	20	21	22	25	25	25	25	25	25
50	25 (42)	33	35	36	37	40	41	44	48	50	50	50	50	50
75		47	51	54	56	58	61	65	69	74	75	75	75	75
100		50 (81)	65	69	75	75	79	83	89	96	100	100	100	100
125			75 (124)	85	89	91	93	102	108	110	119	119	125	125
150				97	104	108	113	120	127	138	144	150	150	150
175				100 (162)	117	120	128	136	143	154	163	172	175	175
200					125 (200)	131	140	149	158	169	184	191	197	200
225						142	152	162	175	189	200	210	220	225
250						145	164	175	192	206	215	228	236	245
275						150 (260)	175	188	205	229	233	242	256	265
300							175 (290)	195	215	235	246	258	275	288
325								199	220	242	256	275	290	304
350								200 (330)	224	245	265	285	305	320
375									225 (360)	248	270	292	310	328
400										250 (390)	275	296	317	335
425											275 (420)	300	320	340
450												300 (450)	322	343
475													324	346
500													325 (485)	348
525														350

The above table is based on 75 mm net depth of application. For other net depth of application, multiply by the factors shown below.

Net depth of application (mm)	25	38	50	63	75	100	125	150	175
Factor	0.77	0.85	0.93	0.98	1.00	1.02	1.04	1.05	1.07

5.9　Change in Groundwater Storage (ΔS)

To estimate the change in groundwater storage, the water levels are observed through a network of observation wells spread over the area. The water levels are highest immediately after monsoon in the month of October or November and lowest just before rainfall in the month of May or June. During the monsoon season, the recharge is more than the extraction; therefore, the change in groundwater storage between the beginning and end of monsoon season indicates the total volume of water added to the groundwater reservoir. While the change in groundwater storage between the beginning and end of the non-monsoon season indicates the total quantity of water withdrawn from groundwater storage. The change in storage (ΔS) is computed as follows:

$$\Delta S = \sum \Delta h \, A \, S_y \qquad \qquad \ldots(16)$$

where,　Δh = change in water table elevation during the given time period;
　　　　A = area influenced by the well; and
　　　　S_y = specific yield.

Groundwater Resource Estimation Committee (1997) recommended that the size of the watershed as a hydrological unit could be of about 100 to 300 sq. km. area and there should be at least three spatially well-distributed observation wells in the unit, or one observation well per 100 sq. km., whichever is more. However, as per IILRI (1974), the following specification may serve as a rough guide:

Size of the Area (ha)	Number of Observation Points	Number of Observation Points per 100 hectares
100	20	20
1,000	40	4
10,000	100	1
1,00,000	300	0.3

The specific yield may be computed from pumping tests. Groundwater Resource Estimation Committee (1997) recommended the following values of specific yield for different geological formations:

(a)　**Alluvial areas**

Sandy alluvium	-	16.0 %
Silty alluvium	-	10.0 %
Clayey alluvium	-	6.0 %

(b)　**Hard rock areas**

Weathered granite, gneiss and schist with low clay content	-	3.0 %
Weathered granite, gneiss and schist with significant clay content	-	1.5 %

Weathered or vesicular, jointed basalt - 2.0 %
Laterite - 2.5 %
Sandstone - 3.0 %
Quartzites - 1.5 %
Limestone - 2.0 %
Karstified limestone - 8.0 %
Phyllites, Shales - 1.5 %
Massive poorly fractured rock - 0.3 %

The values of specific yield in the zone of fluctuation of water table in different parts of the basin can also be approximately determined from the soil classification triangle showing relation between particle size and specific yield (Johnson, 1967).

6.0   ESTABLISHMENT OF RECHARGE COEFFICIENT

Groundwater balance study is a convenient way of establishing the rainfall recharge coefficient, as well as to cross check the accuracy of the various prevalent methods for the estimation of groundwater losses and recharge from other sources. The steps to be followed are:

1.   Divide the year into monsoon and non-monsoon periods.

2.   Estimate all the components of the water balance equation other than rainfall recharge for monsoon period using the available hydrological and meteorological information and employing the prevalent methods for estimation.

3.   Substitute these estimates in the water balance equation and thus calculate the rainfall recharge and hence recharge coefficient (recharge/rainfall ratio). Compare this estimate with those given by various empirical relations valid for the area of study.

4.   For non-monsoon season, estimate all the components of water balance equation including the rainfall recharge which is calculated using recharge coefficient value obtained through the water balance of monsoon period. The rainfall recharge (R_r) will be of very small order in this case. A close balance between the left and right sides of the equation will indicate that the net recharge from all the sources of recharge and discharge has been quantified with a good degree of accuracy.

By quantifying all the inflow and outflow components of a groundwater system, one can determine which particular component has the most significant effect on the groundwater flow regime. Alternatively, a groundwater balance study may be used to compute one unknown component (e.g. the rainfall recharge) of the groundwater balance equation, when all other components are known. The balance study may also serve as a model of the area under study, whereby the effect of change in one component can be used to predict the effect of changes in other components of the groundwater system. In this manner, the study of groundwater balance has a significant role in planning a rational groundwater development of a region.

7.0   CONCLUDING REMARKS

- Water balance approach, essentially a lumped model study, is a viable method of establishing the rainfall recharge coefficient and for evaluating the methods adopted for the quantification of discharge and recharge from other sources. For proper assessment of potential, present use and additional exploitability of water resources at optimal level, a water balance study is necessary. It has been reported that the groundwater resource estimation methodology recommended by Groundwater Resource Estimation Committee (1997) is being used by most of the organisations in India.

- Groundwater exploitation should be such that protection from depletion is provided, protection from pollution is provided, negative ecological effects are reduced to a minimum and economic efficiency of exploitation is attained. Determination of exploitable resources should be based upon hydrological investigations. These investigations logically necessitate use of a mathematical model of groundwater system for analysing and solving the problems. The study of water balance is a pre-requisite for groundwater modelling.

- There is a need for studying unsaturated and saturated flow through weathered and fractured rocks for finding the recharge components from rainfall and from percolation tanks in hard rock groundwater basins. The irrigation return flow under different soils, crops and irrigation practices need to be quantified. Assessment of groundwater quality in many groundwater basins is a task yet to be performed. A hydrological database for groundwater assessment should be established. Also user friendly software should be developed for quick assessment of regional groundwater resources. Tata Infotech Limited (India) has developed a proprietary software 'GEMS' (Groundwater Estimation & Management System) based upon the recommendations of Groundwater Resource Estimation Committee (1997).

- Non-conventional methods for utilisation of water such as through inter-basin transfers, artificial recharge of groundwater and desalination of brackish or sea water as well as traditional water conservation practices like rainwater harvesting, including roof-top rainwater harvesting, need to be practiced to further increase the utilisable water resources.

REFERENCES

1.   Chandra, Satish and R. S. Saksena, 1975. *"Water Balance Study for Estimation of Groundwater Resources"*, Journal of Irrigation and Power, India, October 1975, pp. 443-449.

2.   Central Ground Water Board, 2011. *"Dynamic Ground Water Resources of India (as on 31 March 2009)"*. Ministry of Water Resources, Government of India, November 2011, 225 p.

3.   *"Groundwater Resource Estimation Methodology - 1997"*. Report of the Groundwater Resource Estimation Committee, Ministry of Water Resources, Government of India, New Delhi, June 1997.

4.   IILRI, 1974. *"Drainage Principles and Applications"*, Survey and Investigation, Publication 16, Vol. III.

5.   Johnson, A. I., 1963. *"Application of Laboratory Permeability Data"*, Open File Report, U.S.G.S., Water Resources Division, Denver, Colorado, 34 p.

6.   Johnson, A. I., 1967. *"Specific Yield - Compilation of Specific Yields for Various Materials"*, Water Supply Paper, U.S.G.S., 1962-D, 74 p.

7.   Karanth, K. R., 1987. *"Groundwater Assessment, Development and Management"*, Tata McGraw-Hill Publishing Company Limited, New Delhi, pp. 576-657.

8.   Kumar, C. P. and P. V. Seethapathi, 1988. *"Effect of Additional Surface Irrigation Supply on Groundwater Regime in Upper Ganga Canal Command Area, Part I - Groundwater Balance"*, National Institute of Hydrology, Case Study Report No. CS-10 (Secret/Restricted), 1987-88.

9.   Kumar, C. P. and P. V. Seethapathi, 2002. *"Assessment of Natural Groundwater Recharge in Upper Ganga Canal Command Area"*, Journal of Applied Hydrology, Association of Hydrologists of India, Vol. XV, No. 4, October 2002, pp. 13-20.

10.   Mishra, G. C., 1993. *"Current Status of Methodology for Groundwater Assessment in the Country in Different Region"*, National Institute of Hydrology, Technical Report No. TR-140, 1992-93, 25 p.

11.   *"National Water Policy (2012)"*, Ministry of Water Resources, Government of India.

12.   Sokolov, A. A. and T. G. Chapman, 1974. *"Methods for Water Balance Computations"*, The UNESCO Press, Paris.

13.   Sophocleous, Marios A., 1991. *"Combining the Soilwater Balance and Water-Level Fluctuation Methods to Estimate Natural Groundwater Recharge: Practical Aspects"*. Journal of Hydrology, Vol. 124, pp. 229-241.

14.   Thornthwaite, C. W., 1948. *"An Approach towards a Rational Classification of Climate"*. Geogr. Rev., Vol. 38, No. 1, pp. 55-94.

15.   Thornthwaite, C. W. and J. W. Mather, 1955. *"The Water Balance"*. Publ. Climatol. Lab. Climatol. Drexel Inst. Technol., Vol. 8, No. 1, pp. 1-104.